FLORA OF TROPICAL EAST AFRICA

GYMNOSPERMAE

CYCADACEAE

R. Melville

Trees or shrubs with simple, rarely forked, trunks bearing a crown of leaves or the leaves arising from a tuberous rootstock. Leaves in alternating series of leathery scales and pinnate palm-like foliage leaves. Leaflets with 1 or many parallel nerves, entire or toothed. Cones dioecious, terminal or sub-terminal ; male one or more, with leathery to fleshy, flat or peltate scales bearing on their lower surface numerous crowded one-celled pollen-sacs ; female of flat or peltate scales bearing a large inverted ovule on either side, or consisting of flat blades crowded round the stem apex (*Cycas*) bearing several erect ovules in marginal notches. Seeds large, with a more or less fleshy outer coat and a thin or thick inner shell.

Leaflets 1-nerved with margins entire, coiled in the
    bud ; ovules 1–5 pairs, erect along the margins of
    the blade .   .   .   .   1. **Cycas**
Leaflets many-nerved, usually toothed, straight in the
    bud ; ovules 2, inverted on peltate scales .   .   2. **Encephalartos**

1. CYCAS

L., Sp., Pl. : 1188 (1753)

Shrubs or trees with simple, rarely branched, trunks clothed with woody leaf bases and bearing a crown of pinnate leaves. Leaflets linear, entire, 1-nerved. Scales of the male cones wedge-shaped, often long acuminate, closely imbricate, with ellipsoid pollen-sacs in groups of 3–5 on the lower surface. Female blades crowded round the stem apex, densely woolly at first, later spreading, elongated, widening upwards, with 1–5 pairs of erect ovules in marginal notches. Seeds ellipsoid or globose.

A genus of about 10 species widespread through the tropics and southern subtropics of the old world, but with only one species in Africa.

C. thuarsii *Gaud.* in Freyc., Voy. Aut. Monde 1817–1820, Bot. : 434 (1829); Prain in F.T.A. 6 (2): 345 (1917); U.O.P.Z.: 223 (1949). Type: from Madagascar

Trunk cylindrical, 4–9 m. high, up to 45 cm. diameter. Scale-leaves with deltoid bases 3 cm. wide and linear cusps 3–4 cm. long, 5 mm. wide, externally tomentose, internally smooth, brown. Leaves 1·5–3·0 m. long, 30–60 cm. wide ; median pinnae linear, falcate 22–38 cm. long, 10–20 mm. wide, gradually tapering to the acuminate apex, base decurrent, bright green above, paler below with yellowish green to bright yellow midrib ; rhachis rounded above ; leaflets abruptly replaced by spines on the lower 30–40 cm. of petiole. Male cones cylindrical, yellowish orange, 30–60 cm.

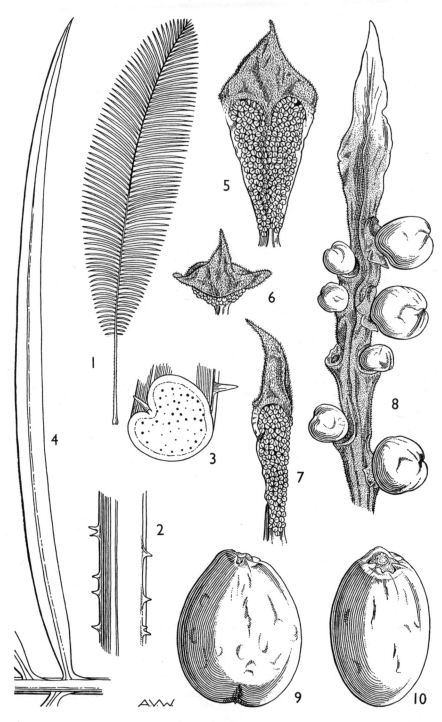

FIG. 1. *CYCAS THUARSII*—**1,** leaf, × 1/18 ; **2,** part of petiole, × 1 ; **3,** T.S. of petiole, × 1; **4,** leaflet, × 2/3 ; **5–7,** male cone-scale from below, front and side, × 2 ; **8,** female frond with young seeds, × 2/3 ; **9, 10,** mature seed from front and side, × 1. 1–3, from plant cult. Kew; 5–7, from *Vaughan* 1158 ; 8, from *Kirk* (coll. 1858) ; 9, 10, from *Thackeray* 1917.

long, 11–20 cm. diameter, median scales spreading horizontally, 3·5–5·0 cm. long, 8–16 mm. wide, gradually tapering to the 2–5 mm. long pedicel, barren apex deltoid, acuminate, brown tomentose, tip upturned. Female blades 15–30 cm. long, linear with expanded tip ovate to lanceolate-acuminate, 7–10 cm. long, 15–30 mm. wide ; ovules 4–5 pairs. Seeds sessile, red, ovoid to subglobose 4·5–6·0 cm. long, 4–6 cm. diameter. Fig. 1.

TANGANYIKA. Tanga, *A.H.* 1158/50 !
ZANZIBAR. Zanzibar Is. : Marahubi, 1 Feb. 1930, *Vaughan* 1158 ! ; Pemba Is. : Mzizima, *Greenway* 2750 !
DISTR. **T3, 6 ; Z ; P ;** Madagascar, Comoro Is., Portuguese East Africa
HAB. In open bushland on sandy loams and coral rocks

SYN. [*C. circinalis* sensu Thouars, Hist. Veg. 2, tab. 1 & 2 (1804), *non* L.]
    C. madagascariensis Miq., Comm. Phyt. 127 (1840). Type : from Madagascar
    C. comorensis Bruant, Cat. Gén. n. 195 : 5 (1888). Type : Comoro Is., *Humblot* (cultivated in France)
    C. circinalis L. subsp. *thuarsii* (Gaud.) Engl., V.E. 2 : 82 (1908)
    C. circinalis L. subsp. *madagascariensis* (Miq.) Schuster in E.P. IV. 1: 73 (1932); T.T.C.L. : 181 (1949)

2. ENCEPHALARTOS

Lehm., Pugill. Pl. 6 : 3 (1834)

Shrubs or trees with erect or prostrate, rarely branched, trunks clothed with woody leaf bases. Leaves, oblong to linear ; leaflets linear-lanceolate to ovate or oblong-lanceolate, with many parallel nerves, spine tipped and more or less spiny toothed, rarely entire, the lowermost often reduced to spines, straight in the bud. Male cones stalked, 1–several, of closely imbricate scales with subpeltate apex (bulla) and a limb closely covered below with ellipsoid pollen sacs. Female cones nearly sessile ; cone-scales with a peltate to subreniform head (bulla), variously faceted and sculptured, bearing a large inverted ovule on either side of the slender quadrangular stalk. Seeds ellipsoid or oblong, often angular by compression.

A genus of about 24 species in tropical and South Africa.

KEY TO FEMALE PLANTS

(All cone-scale characters refer to median scales.)

Angle of inclination of bulla * of cone-scales to pedicel 80°–90° ; height of bulla small (7–10 mm. in dry specimens) :
    Bulla triangular or truncate-triangular ; adaxial margin rounded, obscurely ridged, with sagittal crest ridged to tuberculate ; median leaflets linear-lanceolate with straight tip and 20–24 parallel nerves . . . . . 1. *barteri*
    Bulla rhomboid ; adaxial margin rounded, obscurely ridged with sagittal crest of coarse rounded warts ; median leaflets oblong lanceolate with forward arching tips and 26–43 parallel nerves . . . . . 2. *septentrionalis*
Angle of inclination of bulla of cone-scales to pedicel 40°–60° ; height of bulla considerable (over 10 mm. in dry specimens) :

* For explanation of terms used see Fig. 2, p. 5.

Bulla compressed rhomboidal, truncate-triangular
  above ; sagittal crest of adaxial side consisting of subconical or subulate tubercles ;
  abaxial angle of pedicel ± replaced by a
  ridged and tuberculate facet ; margin of
  median leaflets reflexed, edentate or with 1–3
  teeth near the base above ; indumentum of
  scale-leaves and petiole-bases buff-brown, soft
  and woolly   .   .   .   .   .   .   5. *tegulaneus*
Bulla rhomboid ; sagittal crest of adaxial side
  consisting of short ± cylindrical umbilicate
  tubercles ; all four angles of pedicel acute :
  Adaxial margin of bulla with step-like ridges
    radiating from the sagittal crest ; leaf-
    rhachis with decurrent ridges between the
    leaflets ; indumentum of scale-leaves and
    petiole-bases buff-coloured, closely felted . 4. *bubalinus*
  Adaxial margin of bulla with irregular warts and
    rounded, sometimes umbilicate, tubercles ;
    leaf-rhachis smooth or faintly grooved be-
    tween the leaflets ; mature petiole-bases
    glabrous :
    Lateral margins of the medio-lateral ridge of
      bulla entire .   .   .   .   .   .   3. *hildebrandtii* var.
                                                      hildebrandtii

    Lateral margins of the medio-lateral ridge of
      bulla divaricate dentate   .   .   .   3. *hildebrandtii* var.
                                                      dentatus

KEY TO MALE PLANTS
(All cone-scale characters refer to median scales.)

Bulla of mature cone-scales ± uniformly puberulent ;
  limb spreading ± horizontally :
  Bulla narrow triangular to truncate, sometimes
    rostrate ; median leaflets linear-lanceolate
    with straight tip and 20–24 parallel nerves . 1. *barteri*
  Bulla rhomboid to broad triangular or semi-elliptic ;
    median leaflets oblong lanceolate with forward
    arching tip and 26–43 parallel nerves .   . 2. *septentrionalis*
Bulla of mature cone-scales ± glabrous ; limb
  ascending, spreading, or deflexed :
  Bulla rhomboid to subtriangular :
    Limb of cone-scales ascending at about 60° to the
      cone-axis, oblong, slightly tapering ; leaf-
      rhachis smooth or faintly grooved between
      the leaflets, mature petiole-bases glabrous . 3. *hildebrandtii*
    Limb of cone-scales spreading ± at right angles
      to the cone-axis, broad cuneate ; leaf-
      rhachis with decurrent ridges between the
      leaflets ; indumentum of scale-leaves and
      petiole-bases buff coloured, closely felted . 4. *bubalinus*
  Bulla truncate-triangular :
    Limb of cone-scales deflexed ; bulla with a raised
      quadrangular median facet ; leaf-rhachis dis-
      tinctly grooved between the leaflets ; indu-
      mentum of scale-leaves and petiole-bases
      brown-buff, soft and woolly   .   .   .   5. *tegulaneus*

1. **E. barteri** *Carruth. ex Miq.* in Arch. Neerl. 3 : 243 (1868) ; Prain in Bot. Mag. tab. 8232 (1909) & in F.T.A. 6 (2) : 348 (1917) ; Hutch. & Dalziel, F.W.T.A. 1: 45 (1927); Schuster in E.P. IV. 1: 123 (1932); F.W.T.A., ed. 2, 1 : 32 (1954). Type : Nigeria, Jebba, *Barter* 1692 (K, iso. !)

Trunk 30–150 cm. high, 20–25 cm. diameter. Leaves narrow oblong, tapering gradually to the base, rhachis ± deeply grooved between the leaflets. Median leaflets linear-lanceolate, 10–16 cm. long, 10–15 mm. wide, flexible, tapering from the lower $\frac{1}{3}-\frac{1}{2}$ to a straight, spinescent tip, abruptly contracted to the narrow basal attachment 1·5–4 mm. wide, margins with 0–3–6 spiny teeth ± uniformly spaced, lower surface striate with 20–24 parallel nerves ; lower leaflets lanceolate passing to simple spines ceasing

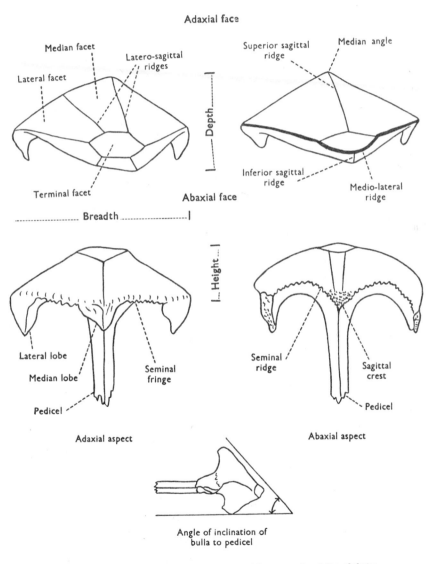

FIG. 2. Terms used in the description of the bulla of the cone-scales of *Encephalartos*.

6–20 cm. above the swollen base. Male cones subcylindrical to fusiform, 8–23 cm. long, 3–5 cm. diameter ; peduncles 8–20 cm. long, 5–9 mm. diameter ; median scales horizontal, 15–20 mm. long tapering to the base, bulla deflexed 12–17 mm. wide, puberulent, triangular to triangular-acuminate-rostrate, truncate, with a rhombic to polygonal terminal facet $\frac{1}{4}$–$\frac{1}{3}$ as wide as the bulla. Female cones oblong ellipsoid, dark olive green, 12–20 cm. long, 8–12 cm. diameter, bulla of median scales deflexed, 4·5–5·6 cm. wide, 19–23 mm. deep, arcuate-triangular, truncate, with the adaxial margin rounded, obscurely ridged and sagittal crest ridged to tuberculate, abaxial face receding, with rounded seminal ridges, terminal facet compressed hexagonal 10–30 mm. wide, lateral lobes triangular to quadrangular, flattened 10–15 mm. long, with lateral facets ± wrinkled and angles acute, entire or ± dentate.

UGANDA. West Nile District : W. Madi, Amua, Sept. 1937, *Eggeling* 3436 ! & W. Madi, Era Reserve, 18 Aug. 1954, *F.D.* 2055
DISTR. U1 ; Ghana, Nigeria, French Sudan and Sudan
HAB. Dry stony hill slopes ; ± 900 m.

2. **E. septentrionalis** *Schweinf.* in Bot. Zeit. 29 : 334 (1871) ; F.T.A. 6 (2) : 350 (1917) ; Schuster in E.P. IV. 1 : 122 (1932) ; F.C.B. 1 : 3, photo. 4 (1948) ; Fl. Pl. Sudan 1 : 1 (1950) ; I.T.U., ed. 2 : 104 (1952) ; Troupin, Fl. Parc Nat. Garamba 1 : 9 & tab. (1956). Type : Sudan, Equatoria Province, Zandeland, Gumango Hill, *Schweinfurth* 2952 (K, iso !)

Trunk 30–200 cm. high and 30 cm. diameter. Leaves narrow oblong, tapering gradually to the base, rhachis distinctly grooved between the leaflets. Median leaflets oblong to lanceolate, ± coriaceous, upper margin ± straight, lower upcurved to the forwardly directed spiny tip, margins with 2–7 spiny teeth often crowded to the base, basal attachment 5–9 mm. wide, lower surface striate with 26–43 parallel nerves ; lower leaflets ovate-lanceolate, spiny, passing to 3–2-furcate and simple spines, ceasing just above the swollen base. Male cones narrow ellipsoid or tapering in the upper half, 12–20 cm. long, 3·0–4·5 cm. diameter, peduncle 10–15 cm. long, 5–12 mm. diameter ; median scales horizontal, triangular, 12–22 mm. long, bulla deflexed, 15–25 mm. wide, 6–10 mm. deep, rhomboid to plano-convex, puberulent, terminal facet lenticular to compressed-rhomboid, $\frac{1}{3}$–$\frac{2}{3}$ as wide as the bulla. Female cones cylindrical, about 23 cm. long, 11 cm. diameter, bulla of median scales deflexed, 4·5–5·0 cm. wide, 2·0–2·7 cm. deep, rhomboid, with the adaxial margin rounded, obscurely ridged, the sagittal crest of coarse, rounded warts, abaxial face receding, seminal ridges distinct, tuberculate dentate, terminal facet rhomboid to semi-elliptic, 14–17 mm. wide, lateral lobes triangular to irregular, 3–6 mm. long with lateral facets warty ridged and the angles tuberculate dentate.

UGANDA. Acholi District : Imatong Mts., Apr. 1938, *Greenway* 3578 ! & Lomwaga Mt., 6 Apr. 1945, *Greenway & Hummel* 7295 ! ; West Nile District : W. Madi, Era Reserve, 18 Aug. 1954, *F.D.* 2055 !
DISTR. U1 ; French Sudan & Sudan, northern Belgian Congo
HAB. Rocky places in grassland and open bushland ; 900–2400 m.

3. **E. hildebrandtii** *A. Br. & Bouché* in Ind. Sem. Hort. Berol. 8 (1874) ; Engler in P.O.A. C : 92 (1895) ; Stapf in Bot. Mag. tabb. 8592–3 (1915) ; Prain in F.T.A. 6 (2) : 351 (1917) ; F.D.O.-A. 1 : 99 (1929) ; Schuster in E.P. IV. 1 : 118 (1932) ; T.S.K. : 1 (1936) ; T.T.C.L. : 181 (1949) ; U.O.P.Z. : 223 (1949). Type : Kenya, Mombasa, *Hildebrandt* (B, holo., K, iso. !)

Trunk to 6 m. high and 30 cm. diameter. Leaves linear-oblanceolate, 2–3 m. long, 60–65 cm. wide, tapering gradually to the base, rhachis rounded above, not or very shallowly grooved between the leaflets. Median leaflets

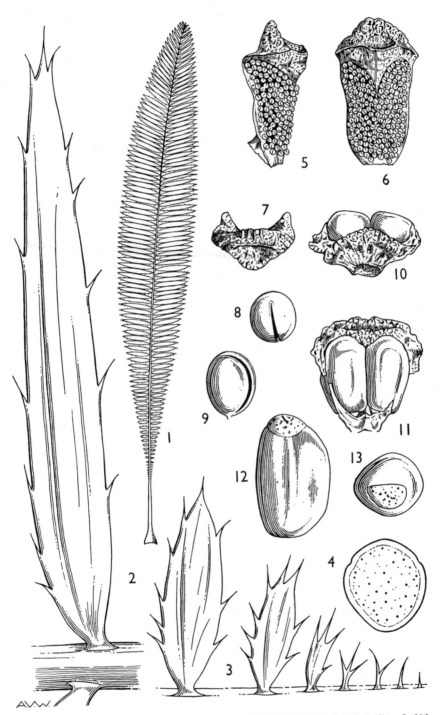

FIG. 3. *ENCEPHALARTOS HILDEBRANDTII* var. *HILDEBRANDTII*—**1**, leaf, × 1/24 ; **2,** 35th leaflet, × 2/3 ; **3,** basal leaflets, left to right 19th, 16th, 12th, 10th, 6th, 4th, 1st, × 2/3 ; **4,** T.S. of petiole, × 1 ; **5–7,** male cone-scale, side, lower and end views, × 1½ ; **8, 9,** pollen sacs, from above and side, × 12 ; **10,** female cone-scale, bulla, × 2/3 ; **11,** female cone-scale, from above, × 2/3 ; **12,** seed, lateral view, × 1 ; **13,** seed showing attachment scar, × 1.

linear-lanceolate, coriaceous, 15–35 cm. long, 13–45 mm. wide, with apex acuminate pungent or 2–3 spiny and 1–4–9 spiny teeth on each margin, often crowded at the base, dark glossy green above, obscurely striate with 26–40 parallel nerves below ; lower leaflets passing from spiny palmate to 3–2-furcate and to simple spines extending to the leaf base. Male cones cylindrical to subconical or fusiform, 20–50 cm. long, 5–9 cm. diameter, greenish to dull red, peduncle 5–25 cm. long, 1·5–3 cm. diameter, with scattered scale-leaves ; median cone-scales ascending, oblong, tapering to the base 20–36 mm. long, with a deflexed compressed rhomboidal to sub-triangular bulla, 19–28 mm. wide, 9–17 mm. deep, adaxial margin rounded, terminal facet rhomboid, $\frac{1}{3}$–$\frac{1}{2}$ as wide and deep as the bulla. Female cones cylindrical, 28–60 cm. long, 15–25 cm. diameter, dull yellow, bulla of median scales deflexed rhomboid, 3·5–5·0 cm. wide, 2·0–3·5 cm. deep, its facets smooth, passing at the margins to irregular warts or ridges and to rounded or umbilicate tubercles, adaxial face with 2 trapezoidal and a median rect-angular facet, terminal facet compressed hexagonal or pentagonal, 12–23 mm. wide, abaxial face receding, subcrescentic with 2 obscure latero-sagittal ridges, lateral lobes triangular to irregular, 7–15 mm. long, with the lateral facets warty-tuberculate and the acute angles ± irregularly dentate.

var. hildebrandtii

Lateral margins of the medio-lateral ridges of the bulla of male and female cones entire. Fig. 3, p. 7.

UGANDA.   Toro District : Mpanza River, *H. H. Johnston* !
KENYA.   Kilifi District : Jilori about 25 km. E. of Malindi, Oct. 1954, *Tweedie* 1230 ! ;
    Lamu District : NE. of Witu, 28 Feb. 1956, *Greenway & Rawlins* 8955 !
TANGANYIKA.   Lushoto District : Oct. 1954, E. Usambara Mts., Sigi, 6 Mar. 1929,
    Greenway 1514 ! & 6 Mar. 1953, *Drummond & Hemsley* 1418 ! ; Tanga District : 8 km.
    SE. of Ngomeni, 31 July 1953, *Drummond & Hemsley* 3583 !
ZANZIBAR.   Zanzibar Is. : Kiwengwa, Jan. 1929, *Greenway* 1250 ! & about 42 km. from
    Zanzibar on Paje road, 5 June 1952, *R. O. Williams* 170 !
DISTR.   U2, K7 ; T3, 6 ; Z
HAB.   Coastal evergreen bushland and lowland forest on red loams and sandy soils
    among gneiss and coral-rocks ; 0–600 m., and about 1200 m. in Uganda

SYN.   [*E. laurentianus* sensu Robyns in F.C.B. 1 : 2 (1948), pro parte, quoad loc.
    ugandensis, et sensu I.T.U., ed. 2, 104 (1952), *non* De Wild.]

    var. dentatus *Melville* in K.B. 1957: 248 (1957). Type: Dar es Salaam, cult., *Wigg*
1044 (K, holo. !)

Lateral margins of the medio-lateral ridges of the bulla of male and female cone-scales acute, divaricate dentate.

DISTR.   Known only from cultivated specimens

4. E. bubalinus *Melville* in K.B. 1957 : 252 (1957). Type : Tanganyika, Masai District, about 20 km. W. of Lake Natron, Moluvane, *Bally* 10600 (K, holo. !)

Trunk to 1·4 m. high and 33 cm. diameter. Leaves oblanceolate, 60–165 cm. long, 20–30 cm. wide, tapering gradually to the base, rhachis with decurrent ridges between the leaflets. Indumentum of scale-leaves and petiole bases buff coloured, closely felted. Median leaflets linear, rigid, coriaceous, 10–20 cm. long, 11–20 mm. wide, upper margin with 2–4 teeth near the base, lower surface obscurely striate with 24–35 parallel nerves ; lower leaflets passing from lanceolate to ovate, with small spiny teeth, to small trifurcate or simple lanceolate and cuneate spines near the petiole base. Male cones ellipsoid to subcylindrical, 11–22 cm. long, 5·5–6·0 cm. diameter, peduncle smooth ; median cone-scales spreading, broad cuneate

23–30 mm. long, 20–25 mm. wide, bulla ± flattened, subtriangular to rhomboid, with a scarcely raised rectangular or triangular median facet 2–6 mm. wide. Female cone not seen ; bulla of median scales deflexed, rhomboid, about 6 cm. wide and 3·6 cm. deep, surface ± uniform, passing at the adaxial margin into low ridges and laterally into small step-like ridges, with the sagittal crest of short, ± cylindrical umbilicate tubercles, seminal fringe of flattened small tubercles irregularly dentate, lateral lobes subtriangular to irregular or flattened, about 10–13 mm. long, lateral facets with short umbilicate tubercles and angles finely tuberculate toothed ; pedicel of scales with four acute angles.

TANGANYIKA. Masai District : between Loliondo and Lake Natron, *Bally* 10600 !
DISTR. **T2** ; endemic to the Nguruman Hills, W. of Lake Natron
HAB. Open bushland on quartzite ridges ; 1300–1600 m.

5. **E. tegulaneus** *Melville* in K.B. 1957 : 249 (1957). Type : Kenya, Mathews Range, Lololokwe, *J. Adamson* (K, holo. !)

Trunk to 7 m. high and 50 cm. diameter. Leaves linear oblanceolate, 120–180 cm. long, 30–40 cm. wide, tapering gradually to the base. rhachis distinctly grooved between the leaflets. Indumentum of scale-leaves and petiole bases buff-brown, softly woolly. Median leaflets oblong lanceolate, 16–22 cm. long, 16–28 mm. wide, rigid, coriaceous, apex pungent, margin reflexed, edentate or with 1–3 teeth on the upper side near the base, ± distinctly striate below with 26–40 parallel nerves ; lower leaflets lanceolate, spiny, passing to ovate-lanceolate, ovate and trifurcate with a few simple spines near the petiole base. Male cones subcylindrical, about 40 cm. long and 13 cm. diameter. with peduncle 20 cm. long and 2·5 cm. diameter, tapering to the base, scales deflexed. Median scales rhomboidal, 45–50 mm. long, bulla arching downwards, truncate triangular, 22–25 mm. wide, with a raised quadrangular median facet 4–8 mm. wide on the adaxial face and a pentagonal to subquadrangular terminal facet 9–13 mm. wide. Female cones, cylindrical, 40 cm. long, 19 cm. diameter. Bulla of median scales deflexed, compressed rhomboidal 5–6 cm. wide, 23–32 mm. deep, ± smooth distally, passing to blunt inclined irregular ridges and blunt flattened tubercles at the adaxial margin, with a sagittal crest of subconical to subulate tubercles. Lateral lobes triangular to quadrangular, 11–17 mm. long, with lateral facets irregularly ridged to warty, inner facets with flattened tubercles and processes and angles bluntly tuberculate toothed ; pedicel of scales quadrangular with angles ± irregularly dentate and the abaxial angle ± replaced by a tuberculate facet.

KENYA. Northern Frontier Province : Mathews Range, Lololokwe, *J. Adamson* !
DISTR. **K1** ; endemic on the Mathews Range
HAB. Bushland on stony slopes at 2100 m.

Imperfectly known species

E. sp. A. Trunk 2 m. or more long, to 30 cm. diameter. Leaves linear lanceolate, about 140–170 cm. long, 24–28 cm. wide, tapering slightly to the rounded apex and ± abruptly to the base, rhachis rounded above, distinctly grooved between the widely spaced leaflets. Median leaflets 3–5 cm. apart, linear to linear lanceolate, 13–15 cm. long, 15–21 mm. wide, coriaceous, pungent, with entire margins or sometimes with 1–2 spiny teeth on the lower margin, lower surface ± distinctly striate with 35–45 parallel nerves ; lower leaflets changing abruptly to ovate lanceolate, ovate followed by 1–2 pairs of spines 17–23 cm. above the swollen base. Male cones about 25 cm. long, young male cones brownish pink with smooth peduncle.

TANGANYIKA. Lushoto District : W. Usambara Mts., World's View, about 1·5 km. W. of Gologolo, 1 Mar. 1953, *Drummond & Hemsley* 1372 !

HAB. Ledges and rocky sides of dry escarpment, at 1800 m.

NOTE. A sterile collection, *Drummond & Hemsley* 3150, from the Kwamshemi—Sakire road, W. Usambara Mts., at 750 m., on rocky outcrops, has the leaflets broader and more closely approximated and the leaves gradually tapering to the base with a gradual transition through 3–2-furcate leaflets to spines.

PODOCARPACEAE

R. MELVILLE

Trees and shrubs with linear to lanceolate or scale-leaves, usually dioecious, the males with small cones or spikes, the females with the cones small or reduced to 1 or 2 fertile scales. Ovules erect or inverted, with the sterile base of the seed scale complex (epimatium) usually ± folded over the ovule and the base of the bracts and cone axis sometimes swelling to form a fleshy receptacle.

PODOCARPUS

Pers., Syn. 2 : 580 (1807)

Shrubs or trees to 30 m. high, with linear lanceolate leaves spirally arranged, or sometimes opposite or whorled on the same tree, dioecious. Male cones cylindrical, catkin-like, shortly stalked; female cones reduced, of 1 or 2 fertile scales bearing solitary inverted ovules and terminating short scaly or leafy lateral branches. Epimatium sometimes fleshy, entirely enclosing the seed, and fused with the thin or thick woody testa. Receptacle sometimes swollen and fleshy.

A genus of about 60 species widespread in the tropics and southern subtropics of both old and new worlds.

Leaves large, up to 10 cm. or more long ; stomata on lower surface only ; torn leaf revealing transverse hypodermal fibres ; opposite leaves on first-year seedlings only ; receptacle of seed more or less swollen :

    Leaves spreading, nearly parallel sided, tapering in upper $\frac{1}{3}-\frac{1}{2}$; receptacle fleshy, red when ripe ; seeds 9–12 mm. long with thin flesh and thin hard shell ; outer bud-scales of terminal buds strap-shaped, acute ; inner with triangular tip and narrow brown band parallel with the margin   .   1. *milanjianus*

    Leaves often pendulous, tapering very gradually from below the middle ; receptacle firm, not colouring ; seeds 24–35 mm. long, with flesh 4–9 mm. thick and no woody shell ; outer bud-scales of terminal buds strap-shaped, with slender linear tip ; inner rounded with short sharp tip and broad brown margin . . . . 2. *ensiculus*

Leaves small, mostly 2–5 cm. long ; stomata on both leaf surfaces ; no hypodermal fibres ; opposite leaves persisting several to many years ; receptacle not swollen :

Seeds commonly ellipsoid, 14–19–23 mm. long,
11–17–21 mm. diameter with woody shell 1–2 mm.
thick ; terminal lobe of middle catkin-scales
1·5 mm. broad ; leaf-apex in adult trees gradually
tapered . . . . 3. *gracilior*
Seeds commonly subglobular, 19–25–35 mm. long,
15·5–22–30 mm. diameter with woody shell
2–6 mm. thick ; terminal lobe of middle catkin-
scales 0·6–1–2 mm. broad ; middle leaves of
shoots of adult trees parallel sided, abruptly
tapered at the tip :
   Seed-shell 3–6 mm. thick ; terminal lobe of catkin-
   scales 0·7–0·9 mm. broad (Usambara Mts.) . 4. *usambarensis*
                                       var. *usambarensis*

   Seed-shell 2–4 mm. thick ; terminal lobe of catkin-
   scales 0·9–1·2 mm. broad (W. of Lake Victoria) 4. *usambarensis*
                                       var. *dawei*

1. **P. milanjianus** *Rendle* in Trans. Linn. Soc., ser. 2, 4 : 61 (1894) ;
Engl. in P.O.A. C : 92 (1895) ; Stapf in F.T.A. 6 (2) : 340 (1917) ;
F.D.O.-A. 1 : 101 (1929) ; T.S.K. : 1 (1936) ; Robyns, F.P.N.A. 1 : 23
(1948) & F.C.B. 1 : 6, tab. 1, (1948) ; Fl. Pl. Sudan 1 : (1950) ; I.T.U.,
ed. 2 : 317, fig. 66a–c. (1952) ; F.W.T.A., ed. 2, 1 : 32 (1954). Type :
Nyasaland, Mt. Mlanje, *Whyte* 34, 39 (BM, syn. !)

A tree to 35 m. high, outer bark of trunk, thin, smooth, brown to grey
brown, exfoliating in papery flakes having a short fracture or with longitudinal fissures 5–15 mm. apart, slash pale brown. Terminal buds of average
branchlets with outer bud scales strap-shaped, acute, intermediate bud
scales with triangular tip and a narrow brown band parallel with the margin.
Leaves spreading, linear, tapering in the upper $\frac{1}{3}$–$\frac{1}{2}$, 2–6–8–15 cm. long,
5–7–9–12 mm. wide, juveniles to 20 cm. long and 13 mm. wide, stomata on
lower side only, fractured surface showing transverse hypodermal fibres.
Male cones solitary or paired, axillary, 15–30–52 mm. long, median scales
with 2 elliptic pollen sacs about 1·8 mm. long, 1·5 mm. across the pair.
Female cones solitary with 1–2 fertile scales, 1–2 seeds maturing. Fruit
green, at first glaucous, obovoid to subglobose, 9–12 mm. long, outer shell
thin, leathery, inner shell thin, woody, fragile; receptacle fleshy, red, 15–
18 mm. long, 13–21 mm. broad, stalk 4–18 mm. long.

UGANDA. Acholi District : Imatong Mts., Lomwaga Mt., 5 Apr. 1945, *Greenway &*
Hummel 7281 ; Toro District : Ruwenzori, Aug. 1938, *Purseglove* 347 ! ; Mbale
District : Elgon, 23 Jan. 1925, *Snowden* 964 !
KENYA. Northern Frontier Province : Mt. Nyiro, 14 Feb. 1947, *J. Adamson* 392 ! ;
Mt. Kenya, *Hutchins* 408 ! ; Londiani District : Tinderet Forest Reserve, 15 June
1949, *Maas Geesteranus* 4984 !
TANGANYIKA. Moshi District : Kilimanjaro, 23 Feb. 1953, *Drummond & Hemsley*
1273 ! ; W. Usambara Mts., Kwebao Forest Reserve, 18 Aug. 1952, *G. R. Williams*
496 ! ; Tukuyu District : Livingstone Forest Reserve, 8 Oct. 1953, *Parry* 233 !
DISTR. U1, 2, 4 ; K1, 3–5 ; T2, 3, 5–7 ; Belgian Congo, Sudan, Nyasaland, Southern
Rhodesia, Northern Rhodesia, Angola, British and French Cameroons
HAB. Upland rain-forest ; 900–3150 m.

SYN. [*P. mannii* sensu Engler in P.O.A. C : 92 (1895), *non* Hook. f.]
     P. uluguruensis Pilger in N.B.G.B. 12 : 82 (1934) ; T.T.C.L. : 454 (1949). Type :
        Tanganyika, Morogoro District, Uluguru Mts., Magali, Aug. 1933, *Schlieben*
        4224
     [*P. latifolius* sensu Peter, F.D.O.-A. 1 : 102 (1929), probab. *non* (Thunb.) Mirb.;
        T.T.C.L.: 454 (1949).]

2. **P. ensiculus** *Melville* in K.B. 1954 : 566 (1954). Type : Tanganyika, W. Usambara Mts., Matondwe Hill, *Drummond & Hemsley* 1340 (K, holo.!)

A tree to 30 m. high, with outer bark of trunk grey brown, 1–3 mm. thick, fibrous, with shallow longitudinal fissures 1–2 cm. apart, slash pinkish red, paling inwards. Terminal buds of average branchlets with outer bud scales strap-shaped, tapering to a linear tip, intermediate bud scales rounded, contracted abruptly into a slender sharp tip and with brown scarious margins. Leaves often pendulous, linear, tapering from below the middle, 6–15 cm. long, 3–8 mm. wide, juveniles to 21 cm. long and 11 mm. wide, stomata on lower surface only, fractured surface showing transverse hypodermal fibres. Male cones solitary, axillary, about 2 cm. long, median scales with 2 narrow elliptic pollen-sacs 2·0–2·2 mm. long, 1·2 mm. across the pair. Female cones solitary, with 1–2 fertile scales, 1 seed maturing. Fruit green to yellowish, ovoid to obliquely ellipsoid, 24–30–35 mm. long, 18–24–28 mm. diameter, shell leathery, 4–9 mm. thick, containing gum cavities to 3 mm. wide; receptacle 4–5 mm. long, hard, stalk 2–5 mm. long.

TANGANYIKA. Morogoro District : Uluguru Mts., Bondwa Hill, 23 Mar. 1953, *Drummond & Hemsley* 1757 !
DISTR. **T**3, 6 ; endemic in Tanganyika
HAB. Upland rain-forest ; 1900–2000 m.

SYN. [*P. ? henkelii* sensu Brenan, T.T.C.L.: 453 (1949), *non* Dallim. & Jacks.]

3. **P. gracilior** *Pilger* in E.P. IV. 5 : 71 (1903) ; Stapf in F.T.A. 6 (2) : 342 (1917) ; T.S.K. : 1 (1936) ; T.T.C.L. : 453 (1949) ; Fl. Pl. Sudan 1 : 1 (1950) ; I.T.U., ed. 2 : 315, fig. 66 d–g, (1952) ; Cuf., Enum. Pl. Aeth. : 1 (1953). Syntypes from Ethiopia and Kenya (B †).

A tree to 30 m. high, outer bark of trunk grey brown to dark brown, exfoliating in rectangular to irregular flakes 2–4–12 cm. long, 1–2–7 cm. wide, slash pink. Terminal buds of average branchlets up to 2 mm. diameter, outer bud scales narrow triangular to strap-shaped, intermediate scales triangular to deltoid with feebly developed point and scarious denticulate margins. Leaves linear to linear-lanceolate, from 10–15 mm. long and 1·5–2·5 mm. broad in senile adults, to 18 cm. long and 6 mm. broad in juveniles, stomata on both surfaces, fractured surface not showing transverse fibres. Male cones 1–3, axillary 10–15–23 mm. long, yellowish brown, terminal lobe of median scales sharply triangular, denticulate, 1·5 mm. broad. Female cones solitary on leafy or scaly branches 7–25 mm. long. Fruit green to yellowish green, ellipsoid, sometimes globose or pyriform 14–19–23 mm. long, 11–17–21 mm. diameter, woody shell 1–2 mm. thick.

UGANDA. Karamoja District : Mt. Debasien, Mareyo, Jan. 1936, *Eggeling* 2704! & Mt. Debasien, 30 May 1939, *A. S. Thomas* 2953! ; Mbale District : Elgon, Kaburon, Jan. 1936, *Eggeling* 2474!
KENYA. Trans-Nzoia District : Elgon, 25 Apr. 1948, *Vesey-Fitzgerald in C. M.* 18617! ; Ravine District : Eldama Ravine, 26 Sept. 1953, *Drummond & Hemsley* 4440! ; Mt. Kenya, NW. side, *Hutchins* 400!
TANGANYIKA. Arusha District : Mt. Meru, 24 Feb. 1953, *Hughes* 160! ; Moshi District : Kilimanjaro, 23 Feb. 1953, *Drummond & Hemsley* 1266! Iringa District : Madibira road at R. Ndembera, Jan. 1954, *Carmichael* 339!
DISTR. **U**1, 3 ; **K**3, 4 ; **T**2, 5, 7 ; Ethiopia, Sudan, Belgian Congo
HAB. Upland rain-forest at 1500–2400 m.

SYN. [*P. elongatus* sensu A. Rich., Tent. Fl. Abyss. 2 : 278 (1851) ; Engl. in P.O.A. C : 92 (1895) ; Peter, F.D.O.-A. 1 : 101 (1929), *non.* (Ait.) L'Hérit.]

4. **P. usambarensis** *Pilger* in E.P. IV. 5 : 70 (1903) ; F.D.O.A. 1 : 101 (1929) ; Stapf in F.T.A. 6 (2) : 341 (1917) ; T.T.C.L.: 454 (1949). Type : Tanganyika, Usambara Mts., Mtai, *Holst* 2467 (K, syn.!)

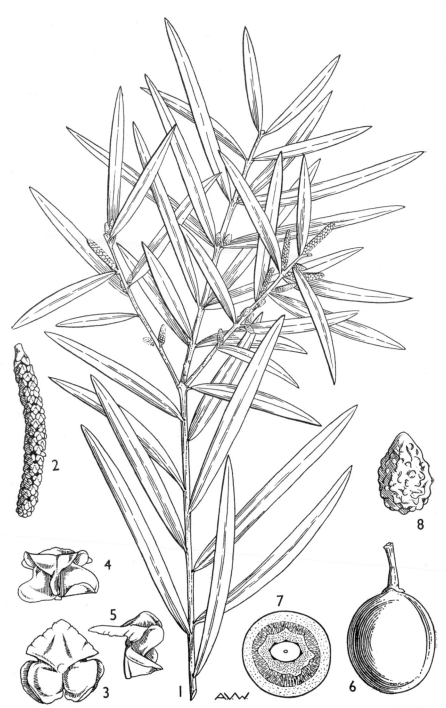

FIG. 4. *PODOCARPUS USAMBARENSIS* var. *DAWEI*, from *Eggeling* 5711—**1**, branch with male catkins,. × 1 ; **2**, male catkin, × 2 ; **3–5**, male catkin-scale from front, below and side, × 12 ; **6**, seed, × 1; **7**, T.S. of seed, × 1 ; **8**, woody stone of seed, × 1.

A tree to 30 m. high, outer bark of trunk grey brown to dark brown, exfoliating in rectangular to irregular flakes 0·5–3 cm. long, 1–2 cm. wide, slash pale brown or pink. Terminal buds of average branchlets 0·7–1·5 mm. diameter, outer bud scales narrow triangular, ± stiff, intermediate bud scales deltoid with a stiff point and margins entire or slightly denticulate. Leaves linear to linear-lanceolate, 10–15 mm. long and 1·5–2·5 mm. broad in senile adults, to 13 cm. long and 8 mm. broad in juveniles, stomata on both sides, fractured surface not showing transverse fibres. Male cones axillary, 4–10–26 mm. long, yellow, terminal lobe of median scales 0·7–1·2 mm. broad, obtusely triangular, scarcely denticulate. Female cones solitary on leafy or scaly branches 7–22 mm. long. Fruit green to yellowish green, subglobular, sometimes ellipsoid, 19–25–35 mm. long, 16–22–30 mm. diameter, woody shell 2–8 mm. thick.

var. usambarensis

Terminal lobe of median scales of male cones 0·7–0·9 mm. broad; woody shell of seeds 3–6–8 mm. thick.

TANGANYIKA. Mbulu District : Mt. Hanang, 5 Feb. 1956, *Greenway* 7602 ! ; Lushoto District : W. Usambara Mts., Shagai Forest, 2 Mar. 1953, *Drummond & Hemsley* 1408 ! & W. Usambara Mts., Jaegertal, 23 Oct. 1952, *Parry* 186 !
DISTR. T2, 3, 5 ; endemic in Tanganyika
HAB. Upland rain-forest at 1500–2000 m.

var. dawei (*Stapf*) *Melville* in K.B. 1954 : 574 (1954). Type: Uganda, Masaka District, S. Buddu, Kagera River, *Dawe* 961 (K, holo. !)

Terminal lobe of median scales of male cones 0·9–1·2 mm. broad; woody shell of seeds 2–4 mm. thick. Fig. 4, p. 13.

UGANDA. Kigezi District : Kayonso Forest, *Fyffe* ! ; Masaka District : Kyebe, 5 Oct. 1953, *Drummond & Hemsley* 4622 ! & Sango Bay, Kaiso Forest, Sept. 1952, *Philip* 511 !
TANGANYIKA. Bukoba District : Minziro Forest, July 1951, *Eggeling* 6248 ! & 11 Aug. 1954, *Willan* 155 !
DISTR. U2, 4 ; T1 ; not known elsewhere
HAB. Swamp-forest at 1150–1400 m.

SYN. [*P. falcatus* sensu Engl. in P.O.A. C : 92 (1895), *non* (Thunb.) Mirb.]
      P. dawei Stapf in F.T.A. 6 (2) : 342 (1917) ; I.T.U., ed. 2 : 315 (1952)

CUPRESSACEAE

R. MELVILLE

Shrubs and trees with scale-like or needle-shaped leaves, monoecious or dioecious. Cone-scales arranged in alternating whorls. Female cones with 1–several ovules at the base of each fertile scale, woody or fleshy and berry-like.

JUNIPERUS

L., Sp. Pl. : 1038 (1753) and Gen. Pl., ed. 5 : 461 (1754)

Shrubs and trees when young or throughout life with linear spiny leaves in whorls of 3 or paired, or when adult with scale-like, opposite leaves or mixed foliage, usually dioecious. Cones on short branchlets or males axillary, of rounded, shield-shaped scales bearing 2–6 pollen sacs; females of 2–4 series of scales in pairs or whorls of 3, each with 1–2 ovules; mature cone fleshy, with few to 1 seed with woody testa.

A genus of about 40 species mainly in Europe and temperate Asia.

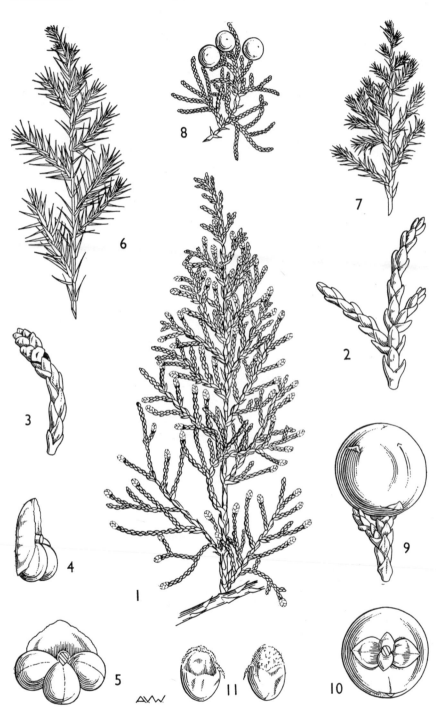

FIG. 5. *JUNIPERUS PROCERA*—**1**, adult foliage with male cones, × 1 ; **2**, adult foliage, × 4 ; **3**, male cone, × 4 ; **4, 5**, male cone-scale, side and back views, × 16 ; **6**, juvenile foliage, × 1 ; **7**, intermediate foliage, × 1 ; **8**, adult foliage and berries, × 1 ; **9**, berry, × 4 ; **10**, berry showing bracts, × 4 ; **11**, seeds, × 4. 1–5, from *Greenway* 7731 ; 6, from *Eggeling* 2476 ; 7, from *Eggeling* 2917 ; 8–11, from *Whyte* 1898A.

1. **J. procera** *Endl.*, Synops. Conif. : 26 (1847) ; Engl. in P.O.A. C : 93 (1895) ; Stapf in F.T.A., 6 (2) : 336 (1917) ; T.S.K. : 3 (1936); F.D.O.-A. : 102 (1938); T.T.C.L.: 177 (1949) ; I.T.U., ed. 2 : 101, fig. 24, photo. 14, (1952) ; Cuf., Enum. Pl. Aeth.: 1 (1953). Type : Ethiopia, Semen, Adda Mariam near Enschedcap, *Schimper* 537 (K, iso. !)

A tree to 40 m. high, outer bark of trunk thin, grey brown, with shallow longitudinal fissures, exfoliating in thin papery strips. Adult branchlets ½–1 mm. diameter with opposite, decussate scale leaves, acute, hooded and keeled at the tip, narrow translucent margins and elliptic oil gland on the back near the base. Juvenile leaves in 3's, linear, spine tipped, decurrent on the branches, not jointed, to 10 mm. long, margins rounded, gland linear, extending ¾ length of leaf and decurrent on stem. Intermediate leaves paired, grading into the adult form. Male cones ellipsoid to subglobose, yellowish, 2–3 mm. long, with 5–6 pairs of rounded obtuse to blunt apiculate scales with 2–3 pollen sacs. Female cones of 3–4 pairs of scales, mature reddish brown to purplish black with blue bloom, subglobose to irregular, 4–6–8 mm. diameter, with 1–4 flattened or triangular seeds. Fig. 5, p. 15.

UGANDA. Karamoja District : Mt. Moroto, Feb. 1936, *Eggeling* 2916 ! 2917 ! ; Mbale District : Elgon, Bukwa, 21 Feb. 1924, *Snowden* 844 ! & Elgon, Sabei, Benet, 12 Dec. 1938, *A. S. Thomas* 2630 !
KENYA. Northern Frontier Province : Furroli, 15 Sept. 1952, *Gillett* 13905 ! Nakuru District : Thomson's Falls, 5 Sept. 1951, *Bogdan* 3233 ! ; Machakos District : Chyulu Hills, 21 May 1938, *Bally* 830 *in C.M.* 7998 !
TANGANYIKA. Mbulu District : Mt. Harang, 11 Feb. 1946, *Greenway* 7695 ! ; Arusha District : Mt. Meru, 18 Jan. 1936, *Greenway* 4424 ! ; Lushoto District : W. Usambara Mts., Lukosi, 23 May 1953, *Drummond & Hemsley* 2687 !
DISTR. U1, 3 ; K1, 3–6 ; T2, 3 ; Ethiopia, Eritrea, Somaliland, Sudan, Belgian Congo, Nyasaland
HAB. Upland dry evergreen forest, often dominant at 1350–3100 m.

INDEX TO GYMNOSPERMAE